For all the amazing teachers in my life - you inspire me every day.

Follow @the.teacher.author for lesson ideas and more!

Published in Australia 2022

Emma's Patterns

Written by JL Cornish

Emma was a little girl,
who loved learning all the time.
She was always learning something new,
in whatever she might find.

And now Emma was learning,
about patterns she could see.
She noticed patterns everywhere,
and how different they could be!

1
2
3
4
5
6
7
8
9
10

One morning in her bedroom,
Emma was making a necklace.
She grabbed some coloured beads,
and started threading them in place.

First she started with a red,
and then added a blue.
She repeated both the colours,
and the pattern continued.

She checked the necklace carefully,
before adding a new bead.
Red, blue, red, blue,
red was the next colour she'd need!

She finished all the beading,
and her necklace looked so neat.
She knew it was a pattern,
because the red and blue repeat!

Or when taking her dog Max,
for a walk around the block,
she noticed something interesting,
about each letterbox.

The first house that she saw,
was the number 33.
The house next door was 35,
was this a pattern she could see?

35
33

The next few houses continued,
37, 39 and 41.
The numbers were increasing,
but they weren't growing by 1.

She kept on walking down the street,
until number 49.
The pattern was increasing,
by 2 numbers at a time!

41
39
37
35
33

Or when she was at school,
she was working with her friend.
They were looking at some different ways,
to make the number 10.

They used double-sided counters,
that were green and red in colour.
If you flipped over a counter,
then it would change to the other.

They organised 10 green counters,
and they flipped over just one.
Now they could see 9 and 1,
but their pattern wasn't done.

They flipped over another counter,
and they noticed something new.
The total was still 10,
but the parts were 8 and 2!

They looked down at the counters,
and wondered what was next.
They thought it would be 7 and 3,
and it was, when they checked!

8+2=10

Emma loved to learn about,
the patterns she could see.
There were patterns all around,
and it filled her heart with glee.

So the next time that you look around,
you might see some patterns too,
and maybe just like Emma,
you can make them continue.

Discussion Questions

In the story, Emma noticed an increasing number pattern. What is the difference between increasing and decreasing patterns?

Emma made a necklace using blue and red beads. She made an AB pattern. What might an ABC pattern look like?

The houses in Emma's street increased by 2 numbers at a time. Can you count forwards and backwards by threes, fives and tens?

Emma made a pattern using combinations to 10. Do you know some more combinations to 10? What are they?

Learning Activities

Emma can see patterns everywhere! Can you make a pattern using claps?

Write down a number pattern and then rub out a number. Show it to someone and see if they know the missing number.

What number is your house? Can you predict what the house next door would be? Write down the number pattern of your street.

Emma made a pattern using combinations to 10. Write down 3 numbers that make 10. Can you write 4 numbers that make 10?

Practise Pages

Practise Pages

Made in United States
North Haven, CT
12 September 2026

18459387R00015